# More Than Just Milk Lactation Science Symposium

# More Than Just Milk Lactation Science Symposium

Editors

**Donna T. Geddes**
**Lisa F. Stinson**
**Sharon L. Perrella**
**Zoya Gridneva**
**Jacki L. McEachran**

MDPI • Basel • Beijing • Wuhan • Barcelona • Belgrade • Manchester • Tokyo • Cluj • Tianjin

*Editors*
Donna T. Geddes
The University of Western Australia
Australia

Lisa F. Stinson
The University of Western Australia
Australia

Sharon L. Perrella
The University of Western Australia
Australia

Zoya Gridneva
The University of Western Australia
Australia

Jacki L. McEachran
The University of Western Australia
Australia

*Editorial Office*
MDPI
St. Alban-Anlage 66
4052 Basel, Switzerland

This is a reprint of Proceedings published online in the open access journal *Proceedings* (ISSN 2504-3900) (available at: https://www.mdpi.com/2504-3900/84/1).

For citation purposes, cite each article independently as indicated on the article page online and as indicated below:

LastName, A.A.; LastName, B.B.; LastName, C.C. Article Title. *Journal Name* **Year**, *Volume Number*, Page Range.

**ISBN 978-3-0365-7180-5 (Hbk)**
**ISBN 978-3-0365-7181-2 (PDF)**

Cover image courtesy of The University of Western Australia

# Contents

*proceedings*

MDPI

*Editorial*

# Statement of Peer Review †

**Jacki L. McEachran *, Zoya Gridneva, Ching-Tat Lai, Lisa F. Stinson, Sharon L. Perrella and Donna T. Geddes**

School of Molecular Sciences, The University of Western Australia, Crawley, WA 6009, Australia
* Correspondence: jacki.mceachran@uwa.edu.au
† Presented at the More Than Just Milk Lactation Science Symposium, Perth, Australia, 25 November 2022.

When submitting conference proceedings to the journal, *Proceedings*, the volume's editors must notify the publisher that they carried out a peer review of all published papers. These reviews were conducted by expert referees, while upholding all of the professional and scientific standards expected of *Proceedings*.

- Type of peer review: single-blind;
- Conference submission management system: email;
- Number of submissions sent for review: 11;
- Number of submissions accepted: 11;
- Acceptance rate (number of submissions accepted/number of submissions received): 100%;
- Average number of reviews per paper: 3;
- Total number of reviewers involved: 3;
- Any additional information on the review process: not applicable.

This publication collates the proceedings of the More Than Just Milk Lactation Science Symposium, held on 25 November 2022 in Perth, Australia.

The symposium was organized by the Geddes Hartmann Human Lactation Research Group, School of Molecular Sciences, The University of Western Australia, Perth, Australia.

The symposium featured keynote speakers from Australia and the United States of America and included student presentations from New Zealand and the United Kingdom. These proceedings provided an opportunity for researchers, students, clinicians, and the community to extend their knowledge and gain a greater understanding of advanced research topics on human lactation, such as breastfeeding experiences, milk composition and production, COVID-19 and human milk, food allergens in human milk, and the impacts of milk on infant growth and development. The symposium is an interactive platform that aims to add diversity, passion, and innovation to lactation research and have a positive impact on mothers, infants, and communities worldwide, showcasing integrative mother/infant-centered research that educates and empowers.

The symposium was well received, with a total of 165 registrations: 22 manuscripts including invited and keynote speakers, were considered; 21 manuscripts were accepted for presentation; and 16 manuscripts were published.

**Citation:** McEachran, J.L.; Gridneva, Z.; Lai, C.-T.; Stinson, L.F.; Perrella, S.L.; Geddes, D.T. Statement of Peer Review. *Proceedings* **2023**, *84*, 1. https://doi.org/10.3390/proceedings2023084001

Published: 10 February 2023



### Organising Committee

Dr. Ching-Tat Lai, The University of Western Australia, Australia
Prof. Donna T. Geddes, The University of Western Australia, Australia
Ms. Jacki L. McEachran, The University of Western Australia, Australia
Dr. Lisa F. Stinson, The University of Western Australia, Australia
Dr. Sharon L. Perrella, The University of Western Australia, Australia
Dr. Zoya Gridneva, The University of Western Australia, Australia

### Scientific Committee

- Dr. Ching-Tat Lai, The University of Western Australia, Australia
- Prof. Donna T. Geddes, The University of Western Australia, Australia

- Dr. Lisa F. Stinson, The University of Western Australia, Australia
- Dr. Sharon L. Perrella, The University of Western Australia, Australia
- Dr. Zoya Gridneva, The University of Western Australia, Australia

**Chairs**

*Session I*

Prof. Donna T. Geddes, The University of Western Australia, Australia
Dr. Sharon L. Perrella, The University of Western Australia, Australia

*Session II*

Dr. Zoya Gridneva, The University of Western Australia, Australia

*Session III*

Prof. Donna T. Geddes, The University of Western Australia, Australia

*Session IV*

Dr. Lisa F. Stinson, The University of Western Australia, Australia

**Keynote Speakers**

Prof. Daniel McAullay, Edith Cowan University, Australia
Dr. Debra J. Palmer, Telethon Kids Institute, The University of Western Australia, Australia
Prof. Donna Geddes, The University of Western Australia, Australia
Prof. Lars Bode, University of California, United States of America
Prof. Yvonne Huack, Curtin University, Australia

**Technical Chair**

Dr. Ching-Tat Lai, The University of Western Australia, Australia

**Registration Chairs**

Ms. Jacki L. McEachran, The University of Western Australia, Australia
Ms. Ashleigh H. Warden, The University of Western Australia, Australia

**Conference Judges—Student Presentations**

Prof. Daniel McAullay, Edith Cowan University, Australia
Dr. Matthew S. Payne, The University of Western Australia, Australia

**Student Prize Winners—sponsored by *Nutrients* (MDPI)**

Frist Prize—Kristin Eu, Curtin University, Australia
Second Prize—Maheshwar Bhasin, The University of Western Australia, Telethon Kids Institute, Australia
Third Prize—Lili Lily Jia, Massey University, New Zealand

**Editors**

Dr. Zoya Gridneva, The University of Western Australia, Australia
Ms. Jacki L. McEachran, The University of Western Australia, Australia
Prof. Donna T. Geddes, The University of Western Australia, Australia

**Proceedings Compilation Team**

Dr. Zoya Gridneva, The University of Western Australia, Australia
Ms. Jacki L. McEachran, The University of Western Australia, Australia
Prof. Donna T. Geddes, The University of Western Australia, Australia

**Conflicts of Interest:** The authors declare no conflict of interest.

proceedings

MDPI

*Abstract*

# Human Milk Microbial Profiles from Allergic Women and Early Childhood Allergy Outcomes †

**Jie Ma [1], Debra J. Palmer [2], Ching Tat Lai [1], Donna T. Geddes [1] and Lisa F. Stinson [1,*]**

[1] School of Molecular Sciences, The University of Western Australia, Crawley, WA 6009, Australia
[2] Telethon Kids Institute, The University of Western Australia, Nedlands, WA 6009, Australia
* Correspondence: lisa.stinson@uwa.edu.au
† Presented at the More Than Just Milk Lactation Science Symposium, Perth, Australia, 25 November 2022.

**Abstract:** Allergic diseases often commence early in childhood and persist into adulthood. An early predictor of an infant/childhood allergy is the infant gut microbiome, which is consistently shaped by the intake of human milk. While numerous studies have reported associations between the infant gut microbiota and allergy outcomes, studies on human milk microbiota are sparse. Here, we explored associations between the human milk microbiota and infant/child allergic sensitisation and diseases. Full-length 16S rRNA gene sequencing was performed on milk samples collected at 3 months postpartum ($n$ = 232) from the Infant Fish Oil Supplementation (IFOS) study. The IFOS cohort consists of infants of allergic mothers, 32.9% of whom went on to develop allergic disease before 3 years of age. An unpaired t-test was used to compare the differences in the milk microbiome with the child allergy outcomes at 1 and 2–3 years of age. The preliminary results showed no difference in the alpha diversity between the children with and without the allergic disease at 1 and 2–3 years of age. However, the relative abundance of several respiratory bacterial species was different in the milk given to infants with and without allergen sensitisation ($p$ = 0.003–0.041): *Streptococcus pneumoniae* was less abundant in the milk fed to infants with allergen sensitisation at 1 year ($p$ = 0.041) and at 2–3 years of age ($p$ = 0.019), whereas *Pseudomonas sp.* was more abundant in the milk fed to infants with allergen sensitisation at 1 year ($p$ = 0.003), but not at 2–3 years of age ($p$ = 0.066). The preliminary analysis identified associations between specific bacteria in human milk and infant/child allergen sensitisation, which require further investigation.

**Keywords:** human milk microbiome; infant/child allergic diseases; infant/child allergen sensitisation

**Citation:** Ma, J.; Palmer, D.J.; Lai, C.T.; Geddes, D.T.; Stinson, L.F. Human Milk Microbial Profiles from Allergic Women and Early Childhood Allergy Outcomes. *Proceedings* **2023**, *84*, 2. https://doi.org/10.3390/proceedings2023084002

Academic Editor: Sharon L. Perrella

Published: 14 February 2023



**Author Contributions:** Conceptualization, D.J.P., L.F.S. and D.T.G.; Writing—original draft preparation, J.M.; writing—review and editing, L.F.S., D.T.G., D.J.P.; experiment and data analysis, J.M.; supervision, L.F.S., D.T.G., D.J.P., C.T.L. All authors have read and agreed to the published version of the manuscript.

**Funding:** J.M., L.F.S., C.T.L., and D.T.G. are supported by an unrestricted research grant from Medela AG, administered by The University of Western Australia. J.M. was supported by an additional SIRF (Scholarships for International Research Fees) scholarship from The University of Western Australia and Ad Hoc Postgraduate Scholarship from Medela AG.

**Institutional Review Board Statement:** Not applicable.

**Informed Consent Statement:** Not applicable.

**Data Availability Statement:** Data sharing not applicable.

**Conflicts of Interest:** J.M., L.F.S., C.T.L., and D.T.G. are supported by an unrestricted research grant from Medela AG, administered by The University of Western Australia. This funding body had no input into manuscript design, data interpretation, or decision to publish.

proceedings

*Abstract*

# Recolonization of Pasteurized Donor Milk with Mother's Own Microbiome †

**Lisa F. Stinson *, Jie Ma and Donna T. Geddes**

School of Molecular Sciences, The University of Western Australia, Crawley, WA 6009, Australia
* Correspondence: lisa.stinson@uwa.edu.au
† Presented at the More Than Just Milk Lactation Science Symposium, Perth, Australia, 25 November 2022.

**Abstract:** Preterm infants are often fed heat-pasteurized (HP) donor human milk (DHM), which is void of live microbes. Previous attempts to restore the microbiome of DHM by inoculation with small quantities of mother's own milk (MOM) have been semi-successful. However, the resulting bacterial profiles are only a partial match to the mother's original microbiota potential due to the altered biochemistry of HP DHM. UVC irradiation reduces bacterial load in donor milk to a similar standard as HP, while preserving non-microbial bioactive components. We therefore hypothesized that the efficacy of DHM restoration will be improved using UVC-irradiated DHM compared to HP DHM. DHM batches ($n$ = 3) were divided into two equal aliquots: one for HP, and one for UVC irradiation. Pasteurized DHM was inoculated with fresh MOM ($n$ = 9) at the following v/v ratios: 5% MOM, 10% MOM, 30% MOM. Samples were incubated at 37 °C for 8 hours, with samples taken every 4 hours. Microbiome restoration was assessed using bacterial culture and viability-coupled 16S rRNA gene sequencing. Both pasteurization techniques were successful with no bacterial growth over the course of the experiment. MOM microbiota were able to expand in both UVC and HP DHM, although growth was more rapid in HP DHM. Overall, HP DHM inoculated with 10–30% MOM and incubated for 4 h most closely resembled baseline fresh MOM. Notably, after 8 hours of incubation, bacterial growth far surpassed baseline MOM levels. This kind of florid growth may be undesirable in a NICU setting where immature and vulnerable infants are fed recolonized DHM. Our results suggest that DHM can be personalized by inoculating with 10–30% MOM and incubating for 4 h. UVC irradiation does not improve recolonization, potentially due to the retention of antimicrobial properties in this type of milk.

**Keywords:** human milk; mother's own milk; donor human milk; human milk microbiome; heat pasteurization; UVC irradiation; bacterial load

**Citation:** Stinson, L.F.; Ma, J.; Geddes, D.T. Recolonization of Pasteurized Donor Milk with Mother's Own Microbiome. *Proceedings* **2023**, *84*, 3. https://doi.org/10.3390/proceedings2023084003

Academic Editor: Sharon L. Perrella

Published: 14 February 2023



**Author Contributions:** Conceptualization, L.F.S. and D.T.G.; methodology, L.F.S. and D.T.G.; analysis, L.F.S. and J.M.; investigation, L.F.S. and J.M.; resources, D.T.G.; data curation, L.F.S.; writing—original draft preparation, L.F.S.; writing—review and editing, D.T.G. All authors have read and agreed to the published version of the manuscript.

**Funding:** This research was funded by unrestricted research grant from Medela GA (Switzerland).

**Institutional Review Board Statement:** This study was approved by The University of Western Australia's Human Research Ethics Committee (RA/4/20/4023), in accordance with its ethics review and approval procedures.

**Informed Consent Statement:** Informed consent was obtained from all subjects involved in the study.

**Data Availability Statement:** Data sharing is not applicable.

**Acknowledgments:** The authors acknowledge Matthew Payne for the use of his laboratory facilities.

**Conflicts of Interest:** L.S. and D.G. receive funding from Medela AG, administered through The University of Western Australia. The funders had no role in the design of the study; in the collection, analyses, or interpretation of data; in the writing of the manuscript; or in the decision to publish the results.

proceedings

MDPI

*Abstract*

# Effect of Human Milk Components on Infant Growth and Body Composition †

**Zoya Gridneva [1,*], Isabella Norrish [1], Azhar Sindi [2,3], Vanessa S. Sakalidis [1], Mya Thway Tint [4], Sharon L. Perrella [1], Mark P. Nicol [5] and Donna T. Geddes [1]**

1 School of Molecular Sciences, The University of Western Australia, Crawley, WA 6009, Australia
2 Division of Obstetrics and Gynaecology, The University of Western Australia, Crawley, WA 6009, Australia
3 College of Applied Medical Sciences, Umm Al-Qura University, Makkah 24381-8156, Saudi Arabia
4 Human Development, Singapore Institute for Clinical Sciences, Agency for Science, Technology and Research (A*STAR), Singapore 117609, Singapore
5 School of Biomedical Sciences, The University of Western Australia, Crawley, WA 6009, Australia
* Correspondence: zoya.gridneva@uwa.edu.au

† Presented at the More Than Just Milk Lactation Science Symposium, Perth, Australia, 25 November 2022.

**Abstract:** Human milk (HM) is an important source of nutrition for infants that provides all elements necessary for their growth and development. Previous studies have reported associations between breastfeeding and a reduced risk of developing obesity and late metabolic diseases, such as type 2 diabetes. However, the underlying mechanisms are poorly understood. Recently, intakes of HM components have been associated with infant body composition (BC), which might be implicated in the reduced risk of childhood obesity among breastfed infants. In this systematic review, electronic bibliographic databases were searched for studies that explored associations between the 24 h intakes of HM macronutrients and bioactive molecules and infant BC and/or growth parameters. Two independent reviewers screened reference lists, extracted and analysed the data and assessed the risk of bias using the National Institute for Clinical Excellence methodological checklist. Of 13 eligible studies, 10 assessed the relationships of infant BC and growth outcomes with HM macronutrients, and eight studies assessed the relationships with HM bioactive components. Significant differential associations with infant anthropometrics and BC were found for intakes and not for concentrations of several HM components, such as lactose, fat and adiponectin. This highlights that measuring concentrations of HM components without quantifying the intake by the infant may be misleading when analysing the relationships of HM components with infant outcomes. Future studies investigating the effect of HM components on infant growth and BC should consider measuring the actual intakes of HM components.

**Keywords:** human milk; breastfeeding; lactation; infant nutrition; infant body composition; infant growth; human milk intake; macronutrient intake; intake of bioactive molecules

**Citation:** Gridneva, Z.; Norrish, I.; Sindi, A.; Sakalidis, V.S.; Tint, M.T.; Perrella, S.L.; Nicol, M.P.; Geddes, D.T. Effect of Human Milk Components on Infant Growth and Body Composition. *Proceedings* **2023**, *84*, 4. https://doi.org/10.3390/proceedings2023084004

Academic Editor: Lisa F. Stinson

Published: 14 February 2023



**Author Contributions:** Conceptualization, D.T.G. and Z.G.; methodology, V.S.S., A.S., I.N. and Z.G.; software, V.S.S., A.S. and I.N.; formal analysis, V.S.S., A.S., I.N. and Z.G.; investigation, V.S.S., A.S., I.N. and Z.G.; resources, D.T.G.; data curation, V.S.S., A.S., I.N. and Z.G.; writing—original draft preparation, V.S.S., A.S., I.N. and Z.G.; writing—review and editing, Z.G., M.P.N., D.T.G., S.L.P. and M.T.T.; supervision, Z.G., D.T.G. and M.P.N.; funding acquisition, D.T.G. All authors have read and agreed to the published version of the manuscript.

**Funding:** This research was funded by an unrestricted research grant from Medela GA (Switzerland).

**Institutional Review Board Statement:** Not applicable.

**Informed Consent Statement:** Not applicable.

**Data Availability Statement:** Data sharing not applicable.

**Conflicts of Interest:** D.T.G. declares participation in the Scientific Advisory Board of Medela AG. Z.G., I.N., A.S., V.S.S., S.L.P. and D.T.G. are/were supported by an unrestricted research grant from Medela AG, administered by The University of Western Australia. Umm Al-Qura University, Saudi Arabia provides a PhD scholarship for A.S.

*proceedings*

MDPI

*Abstract*

# Breastfeeding in Public: An International Exploration of Women's Experiences †

**Yvonne L. Hauck [1,*], Lesley Kuliukas [1], Louise Gallagher [2], Vivienne Brady [2], Charlotta Dykes [3] and Christine Rubertsson [3]**

1 Faculty of Health Sciences, Curtin University, Perth, WA 6102, Australia
2 Trinity College Dublin, The University of Dublin, Dublin 2, Ireland
3 Department of Health Sciences, Lund University, 221 00 Lund, Sweden
* Correspondence: y.hauck@curtin.edu.au

† Presented at the More Than Just Milk Lactation Science Symposium, Perth, Australia, 25 November 2022.

**Abstract:** Qualitative evidence has revealed that women face challenges breastfeeding in public. It is important to gain a greater understanding of these challenges, and, also, how women manage breastfeeding in the presence of someone they are uncomfortable with. Insight into what women perceive as helpful when considering whether to breastfeed in public warrants attention. A cross sectional study was conducted with women living in Australia, Ireland, and Sweden who were currently breastfeeding or had breastfed in the previous two years. Data were collected from online platforms. Content analysis of responses confirmed similarities between countries allowing for collaborative negotiation of final themes. Women ranked responses in relation to their importance and frequencies quantified how often each theme was cited. Responses were collected from 10,910 Australian women, 1835 Irish women, and 1520 Swedish women. Ten themes emerged around how women managed having to breastfeed in the presence of someone they were uncomfortable with. The two highest ranked themes were: 'make the effort to be discreet' and 'move to a private location'. Nine themes each emerged around what was challenging and helpful. 'Unwanted attention' ranked highest in Australia and Sweden, whereas 'environment not suitable' ranked number one in Ireland as the most challenging. Having a 'supportive network' ranked most helpful in Australia and Ireland, whereas 'understanding and acceptance of others' was number one in Sweden. Women's experience of breastfeeding in public presented more international similarities than differences. Themes highlight how public education and infrastructure should be prioritized to better support breastfeeding women.

**Keywords:** breastfeeding; public

**Citation:** Hauck, Y.L.; Kuliukas, L.; Gallagher, L.; Brady, V.; Dykes, C.; Rubertsson, C. Breastfeeding in Public: An International Exploration of Women's Experiences. *Proceedings* **2023**, *84*, 5. https://doi.org/10.3390/proceedings2023084005

Academic Editors: Lisa F. Stinson, Sharon L. Perrella and Donna T. Geddes

Published: 14 February 2023



**Author Contributions:** Y.L.H., L.K., L.G. & C.R. contributed to the conception and study protocol. Y.L.H., L.K., L.G., V.B., C.D. & C.R. all contributed to the analysis of data. The draft manuscript was developed by Y.L.H. & L.K. with critical revisions contributed by L.G., V.B., C.D. & C.R. All authors have read and agreed to the published version of the manuscript.

**Funding:** This research received no external funding.

**Institutional Review Board Statement:** Ethical approval to conduct the study was granted by Curtin University Human Research Ethics Committee in Australia (HRE2018–0037), Research and Ethics Committee, School of Nursing and Midwifery, Trinity College in Ireland (COM_35_17/18) and the Advisory Committee for Research Ethics in Health Education Lund University in Sweden (Reference Number 50–18).

**Informed Consent Statement:** Informed consent was obtained from all subjects involved in the study.

**Data Availability Statement:** The datasets generated and analysed during the current study are not publicly available as they are not all in English and were not combined but may be available from the corresponding author at a reasonable request.

**Acknowledgments:** We thank all the women who contributed to the survey.

**Conflicts of Interest:** The authors declare no conflict of interest.

*proceedings*

MDPI

*Abstract*

# Individual Risk Factors Associated with Delayed Secretory Activation in a Cohort of First-Time Mothers in Western Australia †

**Cassandra Cuffe [1,*], Deborah Ireson [2], Joshua R. Lewis [1,3,4], Roslyn Giglia [5] and Therese A. O'Sullivan [1]**

1 Nutrition & Health Innovation Research Institute, School of Medical and Health Sciences, Edith Cowan University, Joondalup, WA 6027, Australia
2 School of Nursing and Midwifery, South West Campus, Edith Cowan University, Bunbury, WA 6230, Australia
3 Medical School, The University of Western Australia, Crawley, WA 6009, Australia
4 Centre for Kidney Research, Children's Hospital at Westmead, School of Public Health, Sydney Medical School, The University of Sydney, Sydney, NSW 2006, Australia
5 Foodbank of Western Australia, Perth, WA 6105, Australia
* Correspondence: c.cuffe@ecu.edu.au
† Presented at the More Than Just Milk Lactation Science Symposium, Perth, Australia, 25 November 2022.

**Abstract:** Infants of mothers who experience a delay in onset of copious breastmilk, or secretory activation (SA), are at risk of suboptimal feeding during infancy and early termination of breastfeeding, even if their mothers intend to exclusively breastfeed. The aim of this secondary data analysis was to identify individual risk factors associated with delayed SA in a cohort of 159 first-time mothers from a subset participating in an ongoing trial being conducted across multiple hospital sites in Western Australia. Low-risk nulliparous pregnant women were enrolled and followed until 1 week after birth. Maternal self-reported delayed SA status (onset $\geq$ 72 h postpartum) was examined in association with potential risk factors across five dimensions: maternal demographics; maternal anthropometry; obstetric and birth outcomes; newborn characteristics; and infant feeding. Any variable that was $P < 0.1$ in univariate analysis was retained for the multivariable analysis and analyzed with potential confounders maternal age and caesarean birth. The rate of delayed SA was 44% ($n$ = 70). In the multivariable adjusted model mothers with gestational diabetes mellitus (GDM) were significantly more likely to experience delayed SA than mothers without GDM after adjusting for all other factors in the model, (OR = 4.35; 95% CI [1.05, 18.06]). Among mothers with GDM, 11 (79%) had delayed SA. Delayed SA was reported by almost half of first-time mothers, and those with GDM more likely to be affected. Additional support for these mothers may help improve breastfeeding rates.

**Keywords:** human milk; breastfeeding; lactation; delayed secretory activation; gestational diabetes mellitus (GDM); formula feeding; nulliparous pregnant women; lactogenesis II

**Citation:** Cuffe, C.; Ireson, D.; Lewis, J.R.; Giglia, R.; O'Sullivan, T.A. Individual Risk Factors Associated with Delayed Secretory Activation in a Cohort of First-Time Mothers in Western Australia. *Proceedings* **2023**, *84*, 6. https://doi.org/10.3390/proceedings2023084006

Academic Editors: Lisa F. Stinson, Sharon L. Perrella and Donna T. Geddes

Published: 14 February 2023



**Author Contributions:** Conceptualization, R.G., C.C. and T.A.O.; methodology, R.G., C.C., D.I., J.R.L. and T.A.O.; formal analysis, C.C.; writing—original draft preparation, C.C.; writing—review and editing, C.C., R.G., D.I., T.A.O. and J.R.L.; supervision, D.I., J.R.L. and T.A.O. All authors have read and agreed to the published version of the manuscript.

**Funding:** Funding for the ongoing larger project was received from the Department of Health Western Australia and the Stan Perron Charitable Foundation.

**Institutional Review Board Statement:** The larger study was conducted in accordance with the Declaration of Helsinki, and approved by the Human Research Ethics Committee at Edith Cowan University (2019-00572-OSULLIVAN, 8 July 2019).

**Informed Consent Statement:** Informed consent was obtained from all subjects.

**Data Availability Statement:** Data used in this analysis are not publicly available, although other investigators may request access to the dataset if a formal request describing their plans is approved by the principal investigators and the relevant ethics approval is in place.

**Acknowledgments:** The authors would like to thank Julie Hill who prepared the data for analysis.

**Conflicts of Interest:** The authors declare no conflict of interest.

*proceedings*

MDPI

*Abstract*

# Changes to Breast Milk Composition following Increased Maternal Sugar and Fat Consumption †

**Ellen Ward [1,*], Ni Yang [2], Beverly S. Muhlhausler [1,3], Gabriela E. Leghi [3], Merryn J. Netting [4], Matthew J. Elmes [2] and Simon C. Langley-Evans [1]**

1 Commonwealth Scientific and Industrial Research Organisation (CSIRO), Adelaide 5000, Australia
2 School of Biosciences, University of Nottingham, Loughborough LE12 5RD, UK
3 School of Agriculture, Food and Wine, University of Adelaide, Adelaide 5064, Australia
4 Women and Kids Theme, South Australian Health and Medical Research Institute (SAHMRI), Adelaide 5000, Australia
* Correspondence: ellen.ward@csiro.au
† Presented at the More Than Just Milk Lactation Science Symposium, Perth, Australia, 25 November 2022.

**Abstract:** Human milk is influenced by maternal habitual diet, yet we do not fully understand the short-term effects of dietary variations on breast milk macronutrient concentrations. This study aimed to determine if increasing sugar and fat consumption would impact breast milk protein, lactose and lipids. Nine mothers who were exclusively breast-feeding consumed three diets; a control, a higher fat diet and a higher sugar diet at least 1 week apart. Breast milk samples were collected hourly and analysed for concentrations of protein, lactose, triglycerides and cholesterol. Breast milk triglycerides responded to both intervention diets with significantly higher concentrations in comparison to the control diet ($p < 0.001$). Cholesterol concentrations increased more in response to the higher sugar diet than the higher fat diet ($p < 0.005$). Lactose concentrations increased in response to the higher fat diet ($p = 0.006$), and protein decreased in response to the higher fat diet ($p = 0.05$). Variations in breast milk composition were observed over the day with triglyceride and cholesterol concentrations highest at the end of day ($p < 0.001$), and lactose and protein concentrations peaking at hour 10 ($p < 0.001$). Manipulating maternal consumption of fat and sugar impacted concentrations of human milk triglycerides, cholesterol, lactose and protein. Fluctuations were also seen in milk macronutrients in response to time of day.

**Keywords:** human milk; diet; triglycerides; protein; lactose; maternal nutrition; lactation

**Citation:** Ward, E.; Yang, N.; Muhlhausler, B.S.; Leghi, G.E.; Netting, M.J.; Elmes, M.J.; Langley-Evans, S.C. Changes to Breast Milk Composition following Increased Maternal Sugar and Fat Consumption. *Proceedings* **2023**, *84*, 7. https://doi.org/10.3390/proceedings2023084007

Academic Editors: Lisa F. Stinson, Sharon L. Perrella and Donna T. Geddes

Published: 14 February 2023



**Author Contributions:** Conceptualization, E.W., B.S.M., M.J.E., M.J.N., G.E.L., and S.C.L.-E., methodology, E.W., and N.Y.; investigation, E.W.; data curation, E.W., and N.Y.; writing, E.W., B.S.M., M.J.E., and S.C.L.-E. All authors have read and agreed to the published version of the manuscript.

**Funding:** The research was funded by the University of Nottingham-University of Adelaide joint PhD programme.

**Institutional Review Board Statement:** The study was conducted in accordance with the Declaration of Helsinki, and approved by the Faculty of Medicine & Health Sciences Research Ethics Committee of The University of Nottingham (protocol code 191-1901, 14 February 2019).

**Informed Consent Statement:** Informed consent was obtained from all subjects involved in the study.

**Data Availability Statement:** Data from this study will be archived in the University of Nottingham Repository (https://nottingham-repository.worktribe.com) on the completion of EW's PhD studies following the 4 January 2024 and will also be available through the University of Nottingham eTheses database (https://eprints.nottingham.ac.uk/etheses/).

**Conflicts of Interest:** The authors declare no conflict of interest.

proceedings

*Abstract*

# COVID-19 and Human Milk: Are We Prepared for the Next Pandemic? †

**Lars Bode**

Department of Pediatrics, Mother-Milk-Infant Center of Research Excellence (MOMI CORE), Human Milk Institute (HMI), University of California, San Diego, CA 92093, USA; lbode@health.ucsd.edu

† Presented at the More Than Just Milk Lactation Science Symposium, Perth, Australia, 25 November 2022.

**Abstract:** For most babies, human milk is considered the best form of early life nutrition with additional immediate and long-term benefits for health and development. During extreme circumstances and crises, it is therefore of utmost importance to monitor and protect the safety of breastfeeding and the use of human milk. At the onset of the recent COVID-19 pandemic, it was critical to rapidly establish rigorous scientific evidence to ensure that the emerging infectious agent SARS-CoV-2 is not transmitted through human milk. As soon as the WHO declared COVID-19 a pandemic on 11 March 2020, we quickly assembled and activated a multidisciplinary team of human milk researchers and virologists and leveraged our existing Human Milk Research Biorepository to recruit lactating women and collect milk samples—literally from day 1 of the pandemic. We and others used RT-qPCR and found that human milk does indeed occasionally contain SARS-CoV-2 viral mRNA. However, a virus is more than a piece of mRNA, which alone is not able to cause the disease. We therefore tested the hypothesis that viral mRNA found in human milk represents an active, replication competent virus. We validated a cell culture SARS-CoV-2 infectivity assay for use in human milk and discovered the following: (i) the presence of SARS-CoV-2 mRNA in human milk of infected women is rare; (ii) The presence of viral RNA is not the same as presence of active, replication-competent virus. In fact, none of the human milk samples from SARS-CoV-2-infected women contained replication-competent virus, including samples that tested positive for viral mRNA by RT-qPCR; and (iii) even if human milk was contaminated with SARS-CoV-2 during pumping and handling, Holder pasteurization, which is commonly used by human milk banks, inactivates the virus in contaminated human milk. In summary, our research has provided scientific evidence that transmission of SARS-CoV-2 from mother to infant through breastfeeding and the use of human milk is highly unlikely. This study was officially published on 19 August 2020, 161 days after the WHO declared COVID-19 a pandemic. Under normal circumstances, this timeline from ideation to publication would be considered remarkably fast, but it wasn't fast enough during a time of crisis. A total of 161 days of uncertainty has led to fear-based confusion, misinformation, and increased the risk of breastfeeding cessation despite the well-documented benefits of human milk and breastfeeding. The current pandemic has uncovered the urgent and immediate need to invest in research that establishes the safety of breastfeeding and human milk at crisis onset. We therefore call on governments, public health agencies, and the scientific community at large to establish a "rapid response task force" that is capable of rapidly and rigorously monitoring and assessing the safety of breastfeeding and human milk at the onset of the next global health crisis.

**Keywords:** breastfeeding; human milk; lactation; COVID-19; pandemic

**Citation:** Bode, L. COVID-19 and Human Milk: Are We Prepared for the Next Pandemic?. *Proceedings* **2023**, *84*, 8. https://doi.org/10.3390/proceedings2023084008

Academic Editors: Lisa F. Stinson, Sharon L. Perrella and Donna T. Geddes

Published: 14 February 2023



**Funding:** L.B. is the UC San Diego Chair of Collaborative Human Milk Research, endowed by a generous gift from the Family Larsson-Rosenquist Foundation, Switzerland.

**Institutional Review Board Statement:** Human Ethics was not required for this abstract.

**Informed Consent Statement:** Informed consent was not obtained for this review of published literature.

**Data Availability Statement:** Data are not available for this abstract.

**Conflicts of Interest:** The author declares no conflict of interest.

*proceedings*

MDPI

*Abstract*

# The Importance of Human Milk Fatty Acids in Infant Growth and Development—Concentration vs. Relative Abundance vs. Intake †

**Alexandra D. George [1,*], Melvin C. L. Gay [2], Mary E. Wlodek [2,3], Kevin Murray [4] and Donna T. Geddes [2]**

1 Metabolomics Laboratory, Baker Heart and Diabetes Institute, Melbourne 3004, Australia
2 School of Molecular Sciences, The University of Western Australia, Crawley 6009, Australia
3 Department of Obstetrics and Gynaecology, Faculty of Medicine, Dentistry and Health Sciences, The University of Melbourne, Melbourne 3010, Australia
4 School of Population and Global Health, The University of Western Australia, Nedlands 6009, Australia
* Correspondence: alexandra.george@baker.edu.au
† Presented at the More Than Milk Lactation Science Symposium, Perth, Australia, 25 November 2022.

**Abstract:** Fatty acids are the building blocks of the entire human milk lipidome; yet, despite years of research, there is little understanding of their importance in early life health and disease. This may be partly due to the way in measurements have been made of fatty acids in the past. Monthly longitudinal human milk samples were collected from eighteen Western Australian mother–infant dyads, during six months of exclusive breastfeeding ($n$ = 704). Additionally, data including monthly anthropometric measurements, health information, and basic maternal food frequency questionnaires were also collected. Infant 24 h milk intake and total lipid intake were measured at three months. Gas chromatography-mass spectrometry was used to profile human milk fatty acids. Linear regression and Pearson's correlation were used to identify associations between human milk fatty acid composition, intake, maternal characteristics, and infant outcomes. Significant variation in human milk fatty acid composition was identified between dyads, and throughout lactation. Mean infant intake for total lipids was 29.7 ± 9.4 g/day. Intake of many fatty acids, including long chain fatty acids, such as C15:0, C18:1, and C18:2, were positively associated with infant growth ($p < 0.001$). This study identified important findings for many human milk fatty acids not previously described, including C15:0 and C18:2 species. Infant total lipid intake and fatty acid intake perform essential roles in infant growth and development. This study highlights the importance of human milk sampling, analytical methods, and estimating infant intake in relation to infant outcomes. Indeed, these factors should be carefully considered in future research.

**Keywords:** lipids; breastfeeding; human milk; infant nutrition; omics

**Citation:** George, A.D.; Gay, M.C.L.; Wlodek, M.E.; Murray, K.; Geddes, D.T. The Importance of Human Milk Fatty Acids in Infant Growth and Development—Concentration vs. Relative Abundance vs. Intake. *Proceedings* **2023**, *84*, 9. https://doi.org/10.3390/proceedings2023084009

Academic Editors: Lisa F. Stinson and Sharon L. Perrella

Published: 20 February 2023



**Author Contributions:** Conceptualization, A.D.G. and D.T.G.; formal analysis, A.D.G., M.C.L.G. and K.M.; investigation, A.D.G., M.C.L.G., M.E.W. and D.T.G.; methodology, A.D.G. and M.C.L.G.; supervision, M.C.L.G., M.E.W., K.M. and D.T.G.; writing—original draft, A.D.G.; writing—review and editing, A.D.G., M.E.W., K.M. and D.T.G. All authors have read and agreed to the published version of the manuscript.

**Funding:** Financial support was provided by Medela AG. Medela AG had no role in the study design or interpretation, writing of the manuscript, or the decision to submit the manuscript for publication.

**Institutional Review Board Statement:** This study was conducted according to the guidelines of the Declaration of Helsinki and approved by the Human Ethics Research Office of The University of Western Australia (RA4/20/4023).

**Informed Consent Statement:** Informed consent was obtained from all subjects involved in the study.

**Data Availability Statement:** Data are available to researchers upon reasonable request.

**Conflicts of Interest:** D.T.G. declares that she receives salary from Medela AG and declares participation in the Scientific Advisory Board of Medela AG. All other authors have no conflicts of interest to declare.

proceedings

MDPI

*Abstract*

# Milk as a Biological System †

**Donna T. Geddes**

School of Molecular Sciences, The University of Western Australia, Crawley, WA 6009, Australia; donna.geddes@uwa.edu.au

† Presented at the More Than Just Milk Lactation Science Symposium, Perth, Australia, 25 November 2022.

**Abstract:** The benefits of breastfeeding and human milk are unequivocal for both the lactating woman and breastfeeding infant. Emerging evidence suggests that environmental factors impact milk composition and subsequently the rapidly developing infants. Indeed, human milk can be considered a biological system accentuating the need for holistic analyses where multiple milk components are considered in contrast to the common reductionist approach. This systems' biology approach is further supported by the evolution of milk to include components that provide both nutrition and protection. Contemporary human lactation research is beginning to design studies to understand the programming effects of human milk in the context of the environment, genetics, and lifestyle. These studies are also engaging analyses that address the complexity of milk composition in an attempt to understand milk as a biological system. To date, maternal factors such as birth mode, infant sex, breastfeeding mode, body mass index, pregnancy complications, and maternal age and diet have been related to differences in milk composition. Furthermore, components within in the milk also display relationships such as milk bacteria and human milk oligosaccharides suggesting the possibility of modulation within the mammary gland. In conclusion, human milk is highly dynamic and responsive to environmental factors and as such may be amenable to interventions designed to improve infant health.

**Keywords:** human milk; breastfeeding; lactation; milk

**Citation:** Geddes, D.T. Milk as a Biological System. *Proceedings* **2023**, *84*, 10. https://doi.org/10.3390/proceedings2023084010

Academic Editors: Lisa F. Stinson and Sharon L. Perrella

Published: 20 February 2023



**Funding:** This research was funded by an unrestricted research grant from Medela AG (Switzerland).

**Institutional Review Board Statement:** Not applicable.

**Informed Consent Statement:** Not applicable.

**Data Availability Statement:** Data sharing is not applicable.

**Conflicts of Interest:** D.T.G. declares that she receives a salary from a research grant from Medela AG and declares participation in the Scientific Advisory Board of Medela AG. All other authors receive a salary from an unrestricted research grant from Medela AG.

proceedings

Abstract

# The Effects of Therapeutic Ultrasound on Breastmilk Composition: A Quasi-Experimental Pre-Post Design Study †

Kristin Eu [1], Renee McGregor [1], Stephanie Melanko [1], Aik Ping Tay [1], Adelle McArdle [2], Ching Tai Lai [3], Donna T. Geddes [3], Patricia Gaunt [4] and Leanda McKenna [1,*]

1 Curtin School of Allied Health, Curtin University, Bentley, WA 6102, Australia
2 Monash Rural health, Monash University, Clayton, VIC 3800, Australia
3 School of Molecular Sciences, The University of Western Australia, Crawley, WA 6009, Australia
4 Physiotherapy Department, King Edward Memorial Hospital, Subiaco, WA 6008, Australia
* Correspondence: l.mckenna@curtin.edu.au
† Presented at the More Than Just Milk Lactation Science Symposium, Perth, Australia, 25 November 2022.

**Abstract:** Therapeutic ultrasound (TUS) is the most commonly used physiotherapy treatment for inflammatory conditions of the lactating breast (ICLB) which can affect up to 33% of post-partum women. There is no literature on whether the thermal effect of TUS alters breastmilk composition. The aim of this quasi-experimental study was to determine the effect of therapeutic ultrasound (TUS) on the pre–post-session protein, fat and lactose concentration of breastmilk. During a single session conducted in an enclosed room within the community in Perth, Western Australia, TUS was applied for 10 min to the right breast. Pre- and post-TUS 5 mL expressed breastmilk samples were collected. The main outcome measured was concentration of breastmilk protein (g/L), fat (%) and lactose (g/L) before and after application of TUS. There was a significant increase in breastmilk fat concentration between measures taken pre- and post-TUS (mean difference, 1.36%; 95% CI [0.97, 1.75], $p < 0.001$). There was no significant difference in breastmilk protein (mean difference, −0.64; 95% CI [−1.93, 0.64], $p = 0.328$) or lactose concentration (mean difference, −4.77; 95% CI [−11.57, 2.03], $p = 0.169$) between measures taken pre- and post-TUS. In conclusion, TUS applied to the healthy lactating breast does not adversely affect the protein, fat or lactose concentrations of breastmilk. The results of this study support the use of TUS as a safe treatment option for mothers suffering from ICLB.

**Keywords:** breastfeeding; mastitis; physiotherapy; treatment; breastmilk; lactation; therapeutic ultrasound; breastmilk composition

**Citation:** Eu, K.; McGregor, R.; Melanko, S.; Tay, A.P.; McArdle, A.; Lai, C.T.; Geddes, D.T.; Gaunt, P.; McKenna, L. The Effects of Therapeutic Ultrasound on Breastmilk Composition: A Quasi-Experimental Pre-Post Design Study. *Proceedings* **2023**, *84*, 11. https://doi.org/10.3390/proceedings2023084011

Academic Editors: Lisa F. Stinson and Sharon L. Perrella

Published: 21 February 2023



**Author Contributions:** Conceptualization, K.E., R.M., S.M., A.P.T., A.M., D.T.G., P.G. and L.M.; methodology, K.E., R.M., S.M. and A.P.T.; validation, K.E., R.M., S.M. and A.P.T.; formal analysis, K.E.; resources, D.T.G. and L.M.; data curation, K.E., R.M., S.M. and A.P.T.; writing—original draft preparation, K.E., R.M., S.M. and A.P.T.; writing—review and editing, K.E., R.M., S.M., A.P.T., A.M., D.T.G., P.G. and L.M; supervision, A.M., C.T.L., D.T.G., P.G. and L.M. All authors have read and agreed to the published version of the manuscript.

**Funding:** This research received no external funding.

**Institutional Review Board Statement:** This study was approved by the Australian New Zealand Clinical Trials Registry (ACTRN12621001025820p, 5 August 2021).

**Informed Consent Statement:** Informed consent was obtained from all subjects involved in the study.

**Data Availability Statement:** Data sharing not applicable.

**Acknowledgments:** Angela Jacques is acknowledged for her assistance with statistical analysis.

**Conflicts of Interest:** D.T.G. declares participation in the Scientific Advisory Board of Medela AG. C.T.L. and D.T.G. are/were supported by an unrestricted research grant from Medela AG, administered by The University of Western Australia. Medela AG had no role in the design of the study; in the collection, analyses, or interpretation of data; in the writing of the manuscript; or in the decision to publish the results.

*proceedings*

MDPI

*Abstract*

# Investigating the Association between *Staphylococcus aureus* and the Mastitis Spectrum †

**Grace C. McLoughlin [1,*], Noor-Ul-Huda Ghori [1], Donna T. Geddes [2] and Mark P. Nicol [1]**

[1] School of Biomedical Sciences, The University of Western Australia, Crawley, WA 6009, Australia
[2] School of Molecular Sciences, The University of Western Australia, Crawley, WA 6009, Australia
* Correspondence: grace.mcloughlin@research.uwa.edu.au

† Presented at the More Than Just Milk Lactation Science Symposium, Perth, Australia, 25 November 2022.

**Abstract:** Mastitis is an inflammatory breast condition that encompasses a range of clinical presentations, better known as the mastitis spectrum; however, it is poorly understood. Mastitis is believed to be caused by an infection of the breast by bacterial pathogens, with *Staphylococcus aureus* being the most commonly reported causative agent. Gaining a better understanding of the association between the presence of *S. aureus* in human milk and the mastitis spectrum is crucial for improving our understanding of the aetiology of mastitis. This study aims to investigate the association between the presence of *S. aureus* in human milk and the development of mastitis across the mastitis spectrum. Lactating mothers were recruited and asked to complete a questionnaire relating to their breastfeeding history and breast health. A milk sample was aseptically collected from each breast and cultured to isolate *S. aureus*. Preliminary findings show that 27% of lactating mothers with no breast inflammation (12/44) have *S. aureus* present in their milk, compared to 38% of mothers (8/21) reporting some form of mastitis-associated breast inflammation. Of the 21 mothers that reported breast inflammation, 12 mothers were diagnosed with mastitis and 9 mothers reported blocked ducts, of whom 58% and 0% had milk samples positive for *S. aureus*, respectively. These findings indicate that the development of mastitis may be more complex than the presence of *S. aureus* in human milk and further studies are needed to determine the role and pathogenesis of *S. aureus* across the mastitis spectrum.

**Keywords:** human milk; *Staphylococcus aureus*; mastitis; blocked ducts; culturomics

**Citation:** McLoughlin, G.C.; Ghori, N.-U.-H.; Geddes, D.T.; Nicol, M.P. Investigating the Association between *Staphylococcus aureus* and the Mastitis Spectrum. *Proceedings* **2023**, *84*, 12. https://doi.org/10.3390/proceedings2023084012

Academic Editors: Lisa F. Stinson and Sharon L. Perrella

Published: 21 February 2023



**Author Contributions:** Conceptualisation, G.C.M., D.T.G. and M.P.N.; methodology, G.C.M., D.T.G. and M.P.N.; software, G.C.M.; formal analysis, G.C.M., D.T.G. and M.P.N.; investigation, G.C.M., D.T.G. and M.P.N.; resources, D.T.G. and M.P.N.; data curation, G.C.M.; writing—original draft preparation, G.C.M.; writing—review and editing, G.C.M., N.-U.-H.G., D.T.G. and M.P.N.; supervision, N.-U.-H.G., D.T.G. and M.P.N.; funding acquisition, D.T.G. and M.P.N. All authors have read and agreed to the published version of the manuscript.

**Funding:** This research was supported by an Australian National Health and Medical Research Council investigator grant (APP1174455) to M.P.N.

**Institutional Review Board Statement:** The study was conducted in accordance with the Declaration of Helsinki and approved by the Human Research Ethics Committee of the University of Western Australia (protocol code 2021/ET001087 and date of approval 22 March 2022).

**Informed Consent Statement:** Informed consent was obtained from all subjects involved in the study.

**Data Availability Statement:** Data sharing is not applicable.

**Acknowledgments:** The authors would like to acknowledge Medela GA (Switzerland) for providing collection equipment for this study.

**Conflicts of Interest:** D.T.G. declares participation in the Scientific Advisory Board of Medela AG and is supported by an unrestricted research grant from Medela AG, administered by the University of Western Australia. All other authors declare no conflict of interest.

proceedings

MDPI

*Abstract*

# Early Lactation Outcomes after Pregnancies Complicated by Gestational Diabetes †

**Sharon L. Perrella [1,*], Jacki L. McEachran [1], Alethea Rea [2], Stuart A. Prosser [1,3], Mary E. Wlodek [1,4] and Donna T. Geddes [1]**

1 School of Molecular Sciences, The University of Western Australia, Crawley, WA 6009, Australia
2 Mathematics and Statistics, Murdoch University, Murdoch, WA 6150, Australia
3 Western Obstetrics, Balcatta, WA 6021, Australia
4 Medicine, Dentistry and Health Sciences, The University of Melbourne, Melbourne, VIC 3010, Australia
* Correspondence: sharon.perrella@uwa.edu.au

† Presented at the More Than Just Milk Lactation Science Symposium, Perth, Australia, 25 November 2022.

**Abstract:** Continued breastfeeding attenuates maternal and infant risks associated with gestational diabetes mellitus (GDM), which may be more difficult after a GDM pregnancy, with reports of delayed secretory activation and reduced milk production. The total 24 h milk production cycle is not routinely measured; therefore, it is unclear whether reported low milk production is actual or perceived. We aimed to describe early lactation outcomes and 24 h milk production in women with GDM. Women with GDM-complicated pregnancies recorded early feeding practices. The 24 h cycle of milk production was measured at 3 weeks by weighing infants pre- and post-breastfeeds, as well as breast expression volumes. Electronic scales sensitive to 2 g were used. Low 24 h milk production was classified as <600 mL. For women with GDM ($n$ = 40), the median time to the initiation of breastfeeding was 55 min, and delayed secretory activation ($n$ = 24, 60%) was common. Most women achieved frequent milk removal ($\geq 8 \times 24$ h) in the first 24 h after birth ($n$ = 27, 68%) was at one week ($n$ = 38, 95%) and three weeks postpartum ($n$ = 36, 90%). However, median 24 h milk production was 639 mL (range 100–1220 mL), with low milk production measured in $n$ = 16 (40%, range 100–592 mL). Conclusions: Delayed secretory activation and low milk production are more common in women with GDM despite regular milk removal from the breast, suggesting that altered endocrine pathways inherent to GDM may be implicated.

**Keywords:** gestational diabetes mellitus; human milk; low milk production; secretory activation

**Citation:** Perrella, S.L.; McEachran, J.L.; Rea, A.; Prosser, S.A.; Wlodek, M.E.; Geddes, D.T. Early Lactation Outcomes after Pregnancies Complicated by Gestational Diabetes. *Proceedings* **2023**, *84*, 13. https://doi.org/10.3390/proceedings2023084013

Academic Editor: Lisa F. Stinson

Published: 21 February 2023



**Author Contributions:** Conceptualization, S.L.P., M.E.W. and D.T.G.; methodology, S.L.P., J.L.M., A.R. and D.T.G.; analysis, S.L.P. and A.R.; investigation, S.L.P. and S.A.P.; resources, D.T.G.; data curation, S.L.P. and J.L.M.; writing—original draft preparation, S.L.P.; writing—review and editing, D.T.G. All authors have read and agreed to the published version of the manuscript.

**Funding:** This research was funded by an unrestricted research grant paid by Medela AG to The University of Western Australia.

**Institutional Review Board Statement:** This study was approved by the University of Western Australia's Human Research Ethics Committee (RA/4/20/5657), in accordance with its ethics review and approval procedures.

**Informed Consent Statement:** Informed consent was obtained from all subjects involved in the study.

**Data Availability Statement:** The data presented in this study are available on request from the corresponding author. The data are not publicly available due to privacy restrictions.

**Conflicts of Interest:** D.T.G. declares her participation in the Scientific Advisory Board of Medela AG. S.L.P., J.L.M. and D.T.G. receive salaries from an unrestricted research grant paid by Medela AG, administered through The University of Western Australia. The funders had no role in the design of the study; in the collection, analyses, or interpretation of data; in the writing of the manuscript; or in the decision to publish the results.

proceedings

*Abstract*

# Effect of Pasteurisation Techniques on Phages in Human Milk †

Lisa F. Stinson [1], Donna T. Geddes [1] and Lucy L. Furfaro [2,*]

1 School of Molecular Sciences, The University of Western Australia, Crawley, WA 2009, Australia
2 Division of Obstetrics and Gynaecology, Medical School, The University of Western Australia, Crawley, WA 2009, Australia
* Correspondence: lucy.furfaro@uwa.edu.au; Tel.: +61-08-6488-7969

† Presented at the More Than Just Milk Lactation Science Symposium, Perth, Australia, 25 November 2022.

**Abstract:** Bacteriophages (phages) are viruses that are the natural predators of bacteria and highly abundant in human milk and the infant gut microbiome. However, the effect of pasteurisation on human milk phages is unknown. This study, therefore, assessed the effect of holder pasteurisation (HP) and UV-C irradiation (UV) on exogenous bacteriophages inoculated into human milk. Ten donor human milk samples inoculated with a thermotolerant *Escherichia coli* phage (T4) and a thermosensitive *Staphylococcus aureus* phage (BYJ20) were subjected to HP and UV treatments. We found that UV effectively inactivated both phages (8/10 samples; 80%), however, HP was ineffective against the thermotolerant T4 phages (0/10; 0% inactivated). This is the first study to assess the impact of UV and HP methods on the viability of human milk phages. This pilot data suggests that HP methods used by milk banks likely destroy thermosensitive, but not thermotolerant, phages, with implications for early-life virome and bacterial microbiome assembly in donor milk fed infants.

**Keywords:** bacteriophage; donor milk; holder pasteurisation; human milk; UV-C

**Author Contributions:** L.F.S., D.T.G., and L.L.F. designed research; L.F.S. and L.L.F. conducted research; L.L.F. analysed data and had primary responsibility for final content. All authors have read and agreed to the published version of the manuscript.

**Funding:** L.L.F. was supported by the Raine Medical Research Foundation and Forrest Research Foundation. D.T.G. and L.F.S. were supported by an unrestricted research grant from Medela AG (Switzerland), administered by The University of Western Australia.

**Institutional Review Board Statement:** This study was conducted in accordance with the guidelines of the Declaration of Helsinki and approved by the Human Research Ethics Committee at The University of Western Australia (RA/4/1/2369). All participants provided written informed consent.

**Informed Consent Statement:** Informed consent was obtained from all subjects involved in the study.

**Data Availability Statement:** Not applicable.

**Acknowledgments:** The authors would like to acknowledge Jeremy Barr for providing the T4 phage and thank all of the mothers that participated in this study by donating breastmilk. We acknowledge Matthew Payne for the use of their lab space for this study and are grateful to Sharon Perrella, Zoya Gridneva, and Ashleigh Warden for their assistance in recruiting participants for this study.

**Citation:** Stinson, L.F.; Geddes, D.T.; Furfaro, L.L. Effect of Pasteurisation Techniques on Phages in Human Milk. *Proceedings* **2023**, *84*, 14. https://doi.org/10.3390/proceedings2023084014

Academic Editor: Sharon L. Perrella

Published: 23 February 2023

**Conflicts of Interest:** L.F.S. and D.T.G. are supported by an unrestricted research grant from Medela AG, administered by The University of Western Australia. This funding body had no role in the design of the study, collection/analysis/interpretation of data, writing of the manuscript, or in the decision to publish the results. L.L.F. has no conflicts of interest to disclose.

*proceedings*

MDPI

*Abstract*

# Circadian Variation of Human Milk Hormones and Macronutrients: Implications for Sampling and Analysis Strategies †

**Majed A. Suwaydi [1,2,*], Ching Tat Lai [1], Alethea Rea [3], Mary E. Wlodek [1] and Donna T. Geddes [1]**

1 School of Molecular Sciences, The University of Western Australia, Crawley, WA 6009, Australia
2 Department of Medical Laboratory Technology, College of Applied Medical Sciences, Jazan University, Jazan 45142, Saudi Arabia
3 Mathematics and Statistics, Murdoch University, Murdoch, WA 6150, Australia
* Correspondence: majed.suwaydi@reseach.uwa.edu.au
† Presented at the More Than Just Milk Lactation Science Symposium, Perth, Australia, 25 November 2022.

**Abstract:** The daily variation in human milk (HM) hormones and macronutrients is not well characterised and sample protocols are highly variable between studies. Method: During a 24 h period, the mothers ($n$ = 10) hand-expressed small milk samples, immediately before and after each breastfeeding or expression from each breast. Test-weighing was used to determine the volume of HM consumed in each feed. Concentrations of leptin, adiponectin, insulin, fat, and glucose were measured by variable biochemical assays. A linear mixed model was fitted to eleven outcomes of leptin, adiponectin, insulin, fat, glucose (dose and concentration for each), and feed volume. The explanatory variables considered were a circadian cycle (cosine and sine terms), and for dose and concentration outcomes, an indicator variable for pre- or post-feed. The random effect was a circadian cycle (cosine and sine terms) for each mother. Results: The average infant intake of HM was 1060 mL/day (8 to 20 feeds/day). Pre- and post-feed differences were found in the concentrations of leptin, adiponectin, insulin, and fat ($p < 0.05$). Significant circadian variation across the 24 h period was found in adiponectin concentration, insulin (both concentration and dose), fat dose, glucose (both concentration and dose), and milk volume. Conclusion: These results highlight the importance of establishing standardised and rigorous sampling protocols considering all levels of variations (within-feed or circadian) to provide a better determination of the impact of these components on infant health and development.

**Keywords:** human milk; lactation; breastfeeding; hormones; adipokine; macronutrient; circadian rhythm; variation

**Citation:** Suwaydi, M.A.; Lai, C.T.; Rea, A.; Wlodek, M.E.; Geddes, D.T. Circadian Variation of Human Milk Hormones and Macronutrients: Implications for Sampling and Analysis Strategies. *Proceedings* **2023**, *84*, 15. https://doi.org/10.3390/proceedings2023084015

Academic Editors: Lisa F. Stinson and Sharon L. Perrella

Published: 27 February 2023



**Author Contributions:** Conceptualization, M.A.S., D.T.G. and C.T.L.; data curation and writing—original draft, M.A.S. and C.T.L.; writing—review and editing D.T.G., C.T.L. and M.E.W.; visualization, A.R.; supervision, C.T.L., D.T.G. and M.E.W. All authors have read and agreed to the published version of the manuscript.

**Funding:** This work was supported by Jazan University, Saudi Arabia and Medela AG, Switzerland. M.A.S receives a postgraduate scholarship from Jazan University (Saudi Arabia). D.T.G. and C.T.L. receive salaries from an unrestricted research grant paid by Medela AG (Switzerland) and administered by The University of Western Australia.

**Institutional Review Board Statement:** This study was approved by The University of Western Australia's Human Research Ethics Committee (RA/4/1/4492).

**Informed Consent Statement:** Informed consent was obtained from all participants involved in the study.

**Data Availability Statement:** Data sharing not applicable.

**Conflicts of Interest:** D.T.G. declares participation in the Scientific Advisory Board of Medela AG. M.A.S. received a postgraduate scholarship from Jazan University. D.T.G. and C.T.L. receive salary from an unrestricted research grant paid by Medela AG and administered by The University of Western Australia. The funding bodies had no input into the design of the study; in the collection, analyses, or interpretation of data; in the writing of the manuscript; or in the decision to publish the results.

*proceedings*

Abstract

# Effect of *Saccharomyces cerevisiae* Yeast-Based Supplement on Human Milk Oligosaccharide Concentration and Mothers' Perception of Breast Milk Supply: A Randomized Placebo-Controlled Trial †

**Lili Lily Jia, Louise Brough and Janet L. Weber ***

School of Food and Advanced Technology, College of Sciences, Massey University, Palmerston North 4442, New Zealand; ljia@massey.ac.nz (L.L.J.); l.brough@massey.ac.nz (L.B.)
* Correspondence: j.l.weber@massey.ac.nz
† Presented at the More Than Just Milk Lactation Science Symposium, Perth, Australia, 25 November 2022.

**Abstract:** *Saccharomyces cerevisiae* yeast-based supplement (SCYS) is frequently used as a galactagogue. However, its efficacy has only been evaluated in lactating ruminants and sows, not humans. This study aims to investigate the effect of taking an SCYS on human milk oligosaccharide (HMO) concentration and perceived insufficient milk supply (PIM). A randomized, double-blind, placebo-controlled trial was conducted at Palmerston North, New Zealand from May 2019–July 2021. Breastfeeding women with a term infant aged 1–7 months were randomly assigned to consume an SCYS (5 g/day) or placebo for four weeks. Concentration of 11 HMOs were analyzed by UHPLC with fluorescence detection. Online questionnaires were used to evaluate PIM, postnatal distress, infant feeding status, and self-reported side effects of taking SCYS. In addition, the infants' feeding pattern (duration and frequency of breastfeed) were examined using a 24-hour feeding record. Sixty-eight women completed this study. The SCYS had no effect on individual or total HMO concentration. There were no significant differences in PIM, postnatal distress, or infant feeding pattern between the SCYS and placebo groups. However, more participants in the SCYS group than the placebo group perceived that the intervention increased their milk supply (65% vs. 35%, $p < 0.05$). Women in both groups reported adverse effects, but no participant withdrew from this study. The SCYS has no effect on HMO concentration or PIM. Research is needed to investigate the effect of SCYS on breast milk volume and self-reported indicators of milk supply increase.

**Keywords:** human milk; breastfeeding; lactation; human milk oligosaccharides; perceived insufficient milk supply; *Saccharomyces cerevisiae* yeast-based supplement; brewer's yeast; randomized placebo-controlled trial

**Citation:** Jia, L.L.; Brough, L.; Weber, J.L. Effect of *Saccharomyces cerevisiae* Yeast-Based Supplement on Human Milk Oligosaccharide Concentration and Mothers' Perception of Breast Milk Supply: A Randomized Placebo-Controlled Trial. *Proceedings* **2023**, *84*, 16. https://doi.org/10.3390/proceedings2023084016

Academic Editors: Lisa F. Stinson, Sharon L. Perrella and Donna T. Geddes

Published: 13 March 2023



**Author Contributions:** Conceptualization, L.L.J., L.B. and J.L.W.; methodology, L.L.J., L.B. and J.L.W.; formal analysis, L.L.J., L.B. and J.L.W.; investigation, L.L.J.; writing—original draft preparation, L.L.J.; writing—review and editing, L.L.J., L.B. and J.L.W. All authors have read and agreed to the published version of the manuscript.

**Funding:** This research received no external funding.

**Institutional Review Board Statement:** The study was conducted in accordance with the Declaration of Helsinki, and approved by Massey University Human Ethics Committee: Southern A (SOA 18/80, approved on 26 February 2019).

**Informed Consent Statement:** Informed consent was obtained from all subjects involved in the study.

**Data Availability Statement:** Data sharing not applicable.

**Conflicts of Interest:** The authors declare no conflict of interest.

MDPI
St. Alban-Anlage 66
4052 Basel
Switzerland
Tel. +41 61 683 77 34
Fax +41 61 302 89 18
www.mdpi.com

*Proceedings* Editorial Office
E-mail: proceedings@mdpi.com
www.mdpi.com/journal/proceedings

www.ingramcontent.com/pod-product-compliance
Lightning Source LLC
Chambersburg PA
CBHW041955260926
51850CB00019B/2927